SIMPLE MACHINES

WEDGE

Jeff Barger

Rourke Educational Media
rourkeeducationalmedia.com

Before, During, and After Reading Activities

Before Reading: Building Background Knowledge and Academic Vocabulary

"Before Reading" strategies activate prior knowledge and set a purpose for reading. Before reading a book, it is important to tap into what your child or students already know about the topic. This will help them develop their vocabulary and increase their reading comprehension.

Questions and activities to build background knowledge:

1. *Look at the cover of the book. What will this book be about?*
2. *What do you already know about the topic?*
3. *Let's study the Table of Contents. What will you learn about in the book's chapters?*
4. *What would you like to learn about this topic? Do you think you might learn about it from this book? Why or why not?*

Building Academic Vocabulary

Building academic vocabulary is critical to understanding subject content.
Assist your child or students to gain meaning of the following vocabulary words.

Content Area Vocabulary

Read the list. What do these words mean?

- *complex*
- *effort*
- *force*
- *handle*
- *inclined*
- *narrow*

During Reading: Writing Component

"During Reading" strategies help to make connections, monitor understanding, generate questions, and stay focused.

1. *While reading, write in your reading journal any questions you have or anything you do not understand.*
2. *After completing each chapter, write a summary of the chapter in your reading journal.*
3. *While reading, make connections with the text and write them in your reading journal.*
   a) Text to Self – What does this remind me of in my life? What were my feelings when I read this?
   b) Text to Text – What does this remind me of in another book I've read? How is this different from other books I've read?
   c) Text to World – What does this remind me of in the real world? Have I heard about this before? (News, current events, school, etc....)

After Reading: Comprehension and Extension Activity

"After Reading" strategies provide an opportunity to summarize, question, reflect, discuss, and respond to text. After reading the book, work on the following questions with your child or students to check their level of reading comprehension and content mastery.

1. *How do wedges make work easier? (Summarize)*
2. *Why do we need wedges? (Infer)*
3. *What are two wedges that you would find in a classroom? (Asking Questions)*
4. *What is your favorite food that is shaped like a wedge? (Text-to-Self Connection)*

Extension Activity

Look at the doors in your home and at school. Do any of them have a wedge doorstop? List where you found the doorstops and why you think the door needs to be stopped.

TABLE OF CONTENTS

The Six Simple Machines

Machines can be big. Trains and trucks are large machines. They do huge jobs like carry people. These vehicles also move items such as lumber. Not all machines are big. Some are small and helpful.

We can lift a brick with a crowbar. Making muffins is easy. Use a mixer to stir the batter. We write letters with ballpoint pens. Not all machines are the same size. What is the same? They make work take less **effort**.

The ball, at the end or point of the pen, moves ink from a tube to the paper.

What is a simple machine? A simple machine has few or no moving parts. We use them to make work easier. They have been around for thousands of years. Two or more together will make a **complex**, or compound, machine.

There are six types of simple machines: inclined plane, lever, pulley, screw, wedge, and the wheel and axle.

Why do we use machines? They help us with all kinds of jobs. One simple machine can cut. It can break. It can stop a door in its tracks.

What Is a Wedge?

A slice of pie. Tortilla chips. A block of cheese. All of these have triangular shapes. They could be wedges. However, none of them do work. You need the simple machine that looks like a triangle.

A wedge has one wide end. The other end is **narrow**. One or two sides are slanted. This shape helps the wedge do many different tasks.

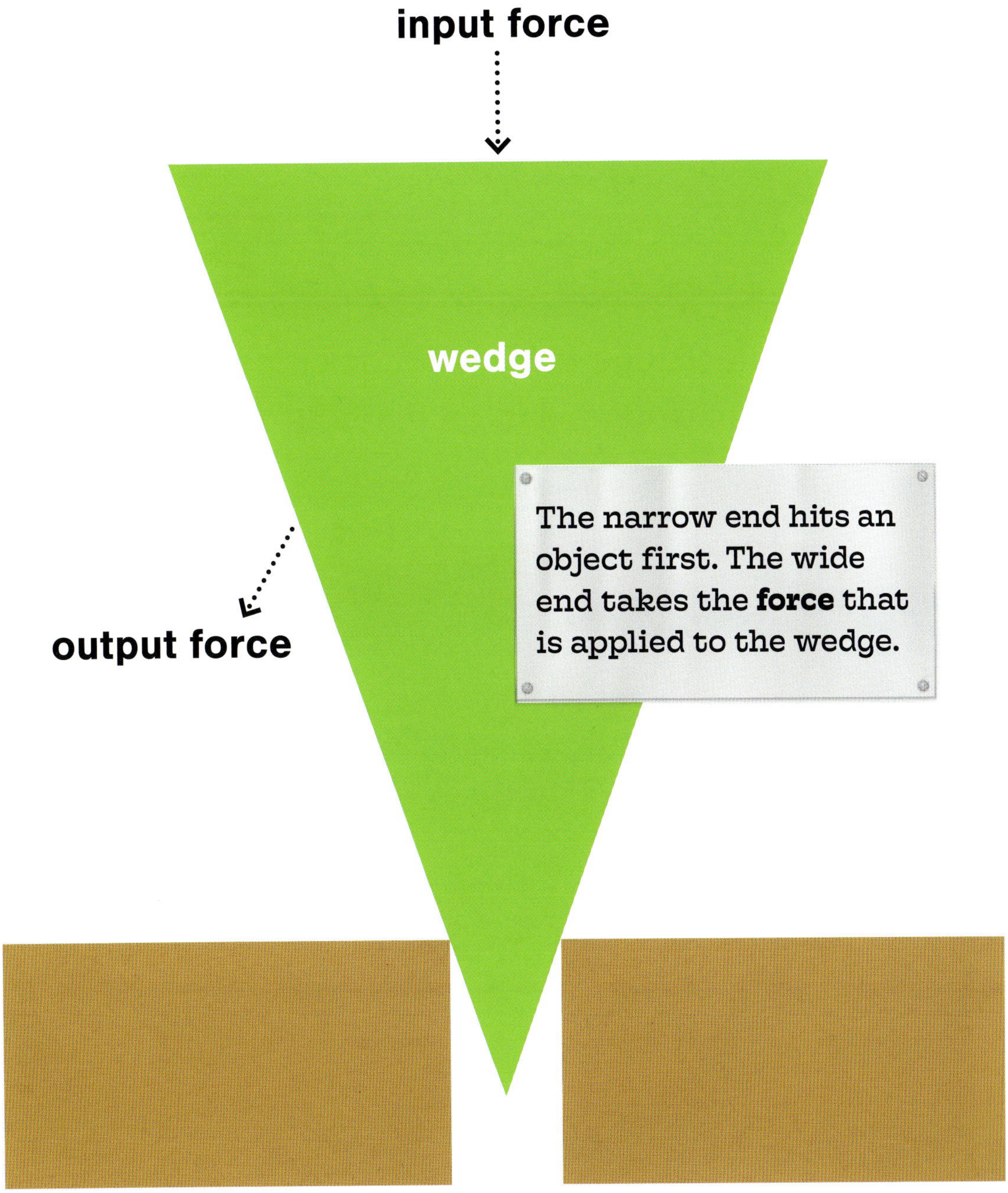

The narrow end hits an object first. The wide end takes the **force** that is applied to the wedge.

Wedges and Food

Can you put a whole pizza in your mouth? Probably not. You need a wedge that cuts. Look closely. A pizza cutter has a thin wedge. It is on the outside circle.

Sometimes pizza can be messy. Pick up a fork. What do you see connected to the **handle**? Four wedges that can break apart food.

If you want to cut your pizza slice in half, you can use a knife. It's also a wedge! Gaze at the edge of the knife. Wedges line the curved side.

Many wedges are used with food. An apple slicer helps with making lunch. Are you going to a wedding? A server is a wedge that will serve delicious cake.

A wedge salad is a block of iceberg lettuce with several toppings.

Working Wedges

Not all wedges go with food. Consider the doorstop. Not very exciting. On the other hand, it is quite useful. The narrow end goes under the door. The wide end blocks the door. This wedge can bring in a cool breeze.

An axe, used to break apart wood, hits a splitting wedge.

An axe is more exciting. It is a wedge attached to a handle. The handle puts force on the blade. This sharp wedge cuts a log in half.

A shovel has a wedge connected to a handle. The wedge slices into the dirt. Speaking of slicing, scissors have two wedges. Each blade joins to a handle.

Wedges can do more than break items apart. They can keep them together. The sharp edge of a nail connects wood. Feeling cold? A zipper will join two sides of your jacket.

A wedge is two **inclined** planes placed back-to-back.

Surprising Wedges

Wedges show up in surprising places. A paper airplane cuts through the wind. The front of a ship, a bow, is wedge shaped. It cuts through water.

The bow is the front of a ship. On the opposite end is the stern.

Look in a mirror. Open your mouth wide. What do you see? Wedges that tear food every day. Keep them clean by brushing and flossing!

Big and small wedges help us work. The scoop on a bulldozer lifts large amounts of dirt. A fork can scoop a piece of pie. Wedges may be simple, but they make life easy.

Activity: Which Wedge Cuts Best?

Which wedge cuts best? Complete this activity to find out. Make sure you have an adult's permission!

Supplies

- cardboard
- plastic or wooden triangle shape
- two or more pieces of bread
- plastic knife
- scissors
- your hand

Directions

1. Using scissors, cut a triangle shape from a piece of cardboard. Each side should be three inches (7.62 centimeters).
2. Lay out two or more pieces of bread side by side.
3. Make a prediction by rating the following items on how well they will cut the bread: cardboard, hand, plastic knife, scissors, triangle. Write down the items in the order of what you think will cut the best. The item that you think will cut best should be listed first.
4. Make small cuts in the bread with each item. Observe how the wedge cuts the bread.
5. After cutting, rate the items again on how well they cut the bread.
6. Compare your predictions with the results of the cutting. How do they compare?

GLOSSARY

complex (KAHM-pleks): having a large number of parts

effort (EF-urt): the activity of trying hard to achieve something

force (fors): any action that produces, stops, or changes the shape or the movement of an object

handle (HAN-duhl): the part of an object that you use to hold, carry, move, or open that object

inclined (in-KLINDE): leaning or slanting

narrow (NAR-oh): measuring a short distance from one side to the other

INDEX

SHOW WHAT YOU KNOW

1. What are the six simple machines?
2. Name two items that are wedges. What do they do?
3. Why would a door need to stay open?
4. What are three wedges that cut?
5. What are the two ends of a ship called?

FURTHER READING

Manolis, Kay, *Wedges*, Bellwether Media, 2016.

Rivera, Andrea, *Wedges*, Abdo Zoom, 2016.

Yasuda, Anita, *Explore Simple Machines!: With 25 Great Projects (Explore Your World)*, Nomad Press, 2011.

ABOUT THE AUTHOR

Jeff Barger is an author, blogger, and literacy specialist. He lives in North Carolina. His favorite wedge is a piece of pizza.

Meet The Author!
www.meetREMauthors.com

www.rourkeeducationalmedia.com

PHOTO CREDITS: Cover and Title Pg ©Da_Ma; Pg 3, 5, 10, 13, 14, 20, 24 ©lolon; Pg 5, 6, 9, 13, 15, 17, 18, ©eriksvoboda; Pg 1, 3, 5, 6, 7, 8,9, 10, 11, 12, 13, 14, 15, 16, 17, 18, 19, 20, 21, 22, 23, 24 ©Amtitus; Pg 4 ©JayLazarin; Pg 5 ©deepblue4you; Pg 6 ©chert61; Pg 8 ©suprun, ©milanfoto; Pg 10 ©shironosov; Pg 11 ©GOLFX; Pg 12 ©koosen; Pg 13 ©Christian Rummel, ©TheCrimsonMonkey; Pg 14 ©jpgfactory, ©Aleks_G; Pg 15 ©SergZSV; Pg 16 ©malerapaso; Pg 17 ©Piotr Polaczyk; Pg 18 ©taborchichakly, ©malerapaso; Pg 19 ©karelnoppe; Pg 20 ©Funwithfood; Pg 21 ©rambo182, ©nongnewnun12; Pg 22 ©vau902

Edited by: Keli Sipperley
Cover and interior design by: Rhea Magaro-Wallace

Library of Congress PCN Data

Wedge / Jeff Barger
(Simple Machines)
ISBN 978-1-64369-043-8 (hard cover)
ISBN 978-1-64369-095-7 (soft cover)
ISBN 978-1-64369-190-9 (e-Book)
Library of Congress Control Number: 2018955992

Rourke Educational Media
Printed in the United States of America,
North Mankato, Minnesota